ercent Hunt

ercents are used in many places. They're used in newspaper ads, food and edicine labels, weather reports, and on tax forms. Sometimes a percent is written ing the word "percent," and sometimes it is written using "%," the symbol for rcent. This symbol is called the **percent sign**.

nd and circle each percent below.

PLUS 33⅓% OFF ALL VITAMINS

110% Price Guarantee

Buy from Uncle Ralph's, and if your purchase is subsequently advertised locally for less, even by us, bring us the ad within 30 days of purchase. We will send you the difference plus 10% of the difference for your trouble.

THEIR PRICE / OUR PRICE
© Uncle Ralph's 1987

NUTRITION INFORMATION PER PORTION

PORTION SIZE (SOLIDS & LIQUID)
PORTIONS PER CONTAINER .... 6½ OUNCES
CALORIES 100
PROTEIN 1
CARBOHYDRATE 18g .... FAT 1g
PERCENTAGE OF U.S. RECOMMENDED DAILY ALLOWANCES (U.S. RDA)
PROTEIN 4g .... SODIUM 920 mg
VITAMIN A 45% .... POTASSIUM 140 mg
VITAMIN C .... THIAMINE .... CALCIUM 15%
CONTAINS LESS THAN 2 PERCENT OF .... RIBOFLAVIN 15%
.... NIACIN 8% .... IRON 4%
NUTRIENTS .... 6% PHOSPHORUS 15%
THE U.S RDA OF THESE
INGREDIENTS: CLAMS, CLAM JUICE (WATER EXTRACT
CLAMS), SALT, SODIUM POLYPHOSPHATE (TO
IN NATURAL .... CALCIUM DISODIUM
TO PRESE...

Neo-Synephrine ½% NASAL SPRAY 2³⁹
NASAL SPRAY adults ½ OUNCE

RUGGED WEAR LTD.
NARRAGANSETT, R.I.
100% COTTON - COLD WATER WASH
NO BLEACH - HANG TO DRY
MADE IN U.S.A.

arena®
85% ANTRON® NYLON
15% LYCRA® SPANDEX
SIZE 34
MADE IN U S A
(CARE OVER)

TV's numbers game

FOR A decade or so, KGO, KPIX and KRON have been losing prime-time viewers to cable TV, independent stations and videocassette recorders, part of a national trend known as "audience erosion." But the erosion may be ending in the Bay Area — and that's good news for San Francisco's three network TV affiliates.

A.C. Nielsen Co. said the three stations captured an average of 68 percent of the viewing audience during prime-time hours (8-11 p.m. Monday through Saturday and 7-11 p.m. Sunday) during the important February ratings period. The figure was identical to that recorded by Nielsen a year ago. Arbitron Ratings Service showed a slight drop, from 69 percent to 67 percent during the year, a difference that may be statistically insignificant.

It's good news to the affiliates, of course, because it means the audience is holding steady — and the more people in front of the set, the more... advertisers.

1450 PETS, SUPPLIES & SERVICES

ENGLISH Springer Pups AKC, Liv/wht, M & F's $200. 686-5290
GERM Shep. Stud svc. Exc hips/temper. Grandfather Nat'l Ch '87. Bidline 50% Tucker Hill/Germ. $500 fee Willie, 661-5039 QUALITY!!!!!
GERMAN Shepherd Pups Pure Bred AKC reg. Champ Sire & Dam. Pets $350. Show quality

Area Rug Cleaning 20% OFF SALE

PRUNE JUICE
A WATER EXTRACT of DRIED PRUNES

...icious, 100% natural fruit juice
...a smooth satisfying taste. Try it
...ed as a healthful breakfast juice,
...utritious between-meal drink.

Appliances aren't **10%**, they aren't **30%**, or even **50%**,

They are 100% of our business
We have to do it right!

30 Year ADJUSTABLE RATE
MORTGAGE LOANS
12% LIFETIME CAP

7.5%
INITIAL RATE

9.539%
A.P.R.

80% Loan to Value
Up to $300,000
1.5% plus $200
FOR INFORMATION OR APPLICATION

GALLUP GRAPHICS
Favorite Exercise Equipment
American-owned equipment

Jogging treadmill	2%
Trampoline	7%
Rowing machine	8%
Exercise bicycle	21%
Weights	26%

■ The Final Four semifinal game telecasts were a hit with the viewers. The first game (Sracuse-Providence) registered a 10.7 rating, 22 percent better than last year's Louisville-LSU game. The nightcap (Indiana-UNLV) got a 13.8 rating, up 29 percent over last year's Duke-Kansas telecast.

CHANTAL SKIN CARE ALOE VERA 72% SKIN CREAM

...88 by Key Curriculum Project, Inc.
...ot duplicate without permission.

1

Percent Means Hundredths

A fraction with a denominator of 100 is easy to write as a percent. Just write down the numerator of the fraction and follow it with a percent sign. The percent sign means hundredths. (It even looks like a "1" and two "0"s.)

$$\frac{25}{100} = 25\% \qquad \text{"25 hundredths = 25 percent"}$$

What part of each large square is shaded? Give each answer as a fraction with denominator 100 and also as a number with a percent sign.

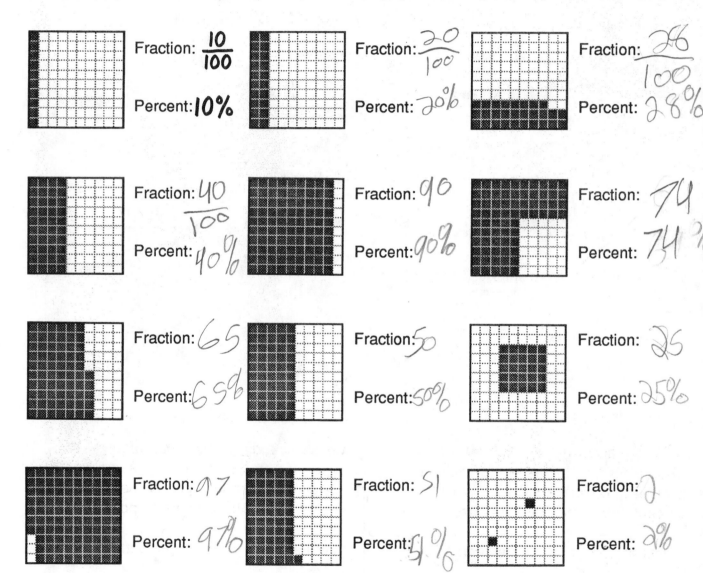

Fraction: $\frac{10}{100}$

Percent: 10%

Fraction: $\frac{20}{100}$

Percent: 20%

Fraction: $\frac{26}{100}$

Percent: 28%

Fraction: $\frac{40}{100}$

Percent: 40%

Fraction: 90

Percent: 90%

Fraction: 74

Percent: 74%

Fraction: 65

Percent: 65%

Fraction: 50

Percent: 50%

Fraction: 25

Percent: 25%

Fraction: 97

Percent: 97%

Fraction: 51

Percent: 51%

Fraction: 2

Percent: 2%

2

Key to
Percents®

Percent Concepts

By Steven Rasmussen and David Rasmussen

Name _____ Class _____

TABLE OF CONTENTS

Cover art by James Dyekman

About the Cover:

Percent is a mathematical concept that has been used since the end of the fifteenth century in business problems such as computing interest, profits and losses, and taxes. However, the idea had its origin much earlier. When the Roman emperor, Caesar Augustus, levied a tax on all goods sold at auction, *centesima rerum venalium*, the rate was 1/100 of the value. Other Roman taxes were 1/20 on the value of every freed slave and 1/25 on the price of every slave sold. Without recognizing percentages as such, the Romans used fractions easily converted to hundredths.

Just as in Roman times, many of the taxes we pay today are based on percents. Sales taxes are based on a percent of the sales price of items we buy. We pay a percent of our incomes as income tax. Property taxes are based on a percent of a property's value.

On the cover of this booklet the Roman emperor, Caesar Augustus, presides over the Roman auction.

Note: Some of this material is from "Percent," an essay by Harlen E. Amundson published in *Topics for the Mathematics Classroom*, the Thirty-Fifth Yearbook of the National Council of Teachers of Mathematics. Used by permission of the NCTM.

Copyright © 1988 by Key Curriculum Project, Inc. All rights reserved.
® *Key to Fractions, Key to Decimals, Key to Percents, Key to Algebra, Key to Geometry, Key to Measurement*, and *Key to Metric Measurement* are registered trademarks of Key Curriculum Press.
Published by Key Curriculum Press, 1150 65th Street, Emeryville, CA 94608
Printed in the United States of America   31 30 29 28 27 26 25   13 12 11 10 09 08 07   ISBN 978-0-913684-57-3

What part of each group is circled? Give each answer as a fraction with denominator 100 and as a number with a percent sign.

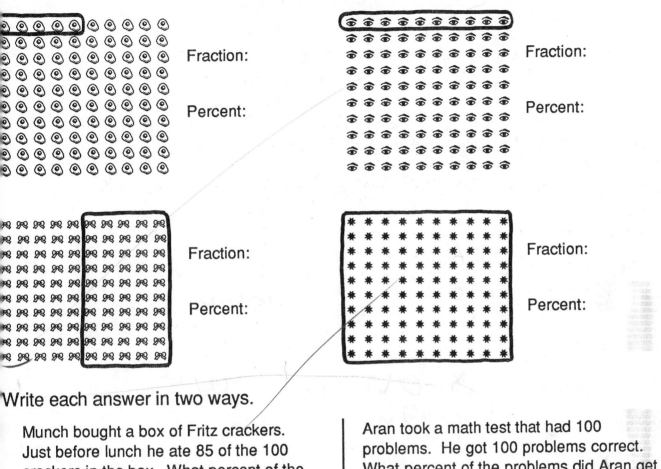

Fraction:

Percent:

Fraction:

Percent:

Fraction:

Percent:

Fraction:

Percent:

Write each answer in two ways.

Munch bought a box of Fritz crackers. Just before lunch he ate 85 of the 100 crackers in the box. What percent of the crackers did Munch eat?

Fraction: 85   Percent: 85%

Aran took a math test that had 100 problems. He got 100 problems correct. What percent of the problems did Aran get correct?

Fraction: 100   Percent: 100%

Anton's teacher gave her class 100 minutes for a math test. Anton took 95 minutes to complete the test. What percent of the time did Anton use?

Fraction: 95   Percent: 95%

Monica is a quality inspector for a supermarket. She checked a sample of 100 oranges for spoilage. None were spoiled. What percent were spoiled?

Fraction: 100   Percent: 100%

The word percent comes from the Latin word "centum" which means one hundred. So does the word "cent." A cent is one hundredth of a dollar. Percent means "out of every 100" or "compared to 100" or "per hundred."

$$50\% \text{ means} \begin{cases} 50 \text{ out of every } 100 \\ 50 \text{ compared to } 100 \\ 50 \text{ per hundred}. \end{cases}$$

Complete each sentence. Use "out of every 100" or "compared to 100" or "per hundred" and a fraction with 100 as denominator.

12% means: __12 out of every 100__ or $\frac{12}{100}$.

1% means: __1 out of every 100__ or $\frac{1}{100}$.

18% means: __18 out of every 100__ or $\frac{18}{100}$.

6% means: __6 out of every 100__ or $\frac{6}{100}$.

50% means: __50 out of every 100__ or $\frac{5}{?}$.

98% means: __98 out of every 100__ or $\frac{98}{100}$.

100% means: __100 out of 100__ or 1.

Write each percent in three ways.

Using words	As a fraction with denominator 100	As a number with a percent sign
36 out of every 100	$\frac{36}{100}$	**36%**
48 out of every 100	$\frac{48}{100}$	46
3 out of every 100	$\frac{3}{100}$	3%
7 out of every 100	$\frac{7}{100}$	7%
99 compared to 100	$\frac{99}{100}$	99%
100 out of 100	1	100%

Each large square below is divided into _____ small equal squares.

Shade 25%.   Shade 50%.   Shade 75%.   Shade 100%.

Shade 1%.   Shade 10%.   Shade 11%.   Shade 99%.

Each group of stars below has _____ stars.

Circle 15%.   Circle 60%.   Circle 95%.

Each rectangle below is divided into _____ equal parts.

Shade 33%.   Shade 90%.   Shade 0%.

_____% is not shaded.   _____% is not shaded.   _____% is not shaded.

Answer each question using a complete sentence.

90% of the songs played on Station KRAZ are rock and roll. Yesterday afternoon the DJ played 100 songs. How many were rock and roll?

90 songs were rock and roll.

Sui bought a sheet of 100 postage stamps. She used 82% of them to mail greeting cards. How many stamps did she have left?

All and None as Percents

Shade **all** of the square.

What percent of the
square is shaded? _____%

All of something is _____% of it.

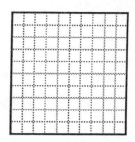

$$1 = \underline{\hspace{2cm}}\%$$

Shade **none** of the square.

What percent of the
square is shaded? _____%

None of something is _____% of it.

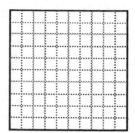

$$0 = \underline{\hspace{2cm}}\%$$

Circle all of the stars.

Shade 100%.

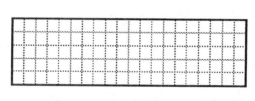

Circle 100% of the dots.

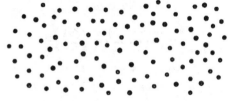

Circle none of the stars.

Shade 0%.

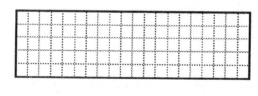

Circle 0% of the dots.

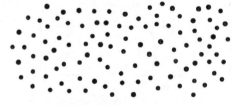

Answer each question using a complete sentence.

On the day of the Mr. Mean's class picnic, everyone came to school. What percent of the class was present?

_____% of the class was present.

On the next day Mr. Mean gave a math test. No one showed up. What percent of the class was present?

You can use 100% and 0% even when you can't divide something into hundredths. % of any number is zero (none of it). 100% of any number is the number (all of it).

Shade 100%.

Shade 0%.

Shade 100%.

Circle 0%.

Circle 100%.

Circle 0%.

0% of 8 is __O__.

0% of 25 is _____.

0% of 99 is _____.

100% of 8 is __8__.

100% of 25 is _____.

100% of 99 is _____.

_____ is 100% of 12.

0% of 32 is _____.

_____ is 0% of 64.

0 is _____% of 16.

84 is _____% of 84.

_____% of 1320 is 1320.

35 is 100% of _____.

0 is 100% of _____.

0 is 0% of _____.

se 0% and 100% with amounts of money just as you did with numbers.

0% of $50 is __$O__.

100% of $18.50 is _____.

100% of $100 is _____.

_____ is 0% of $5.

100% of $1 is _____.

_____ is 100% of 25¢.

nswer each question using a complete sentence.

Bill had 30 minutes to complete his math test. He used 100% of his time. How long did he spend on his test?

He spent ___ minutes on his test.

The Stingy Pay Plant workers earn $6.50 per hour. The owner offered them a 0% raise. How much per hour was the raise?

One Half and One Fourth as Percents

Shade **one half** of the square.

What percent of the
square is shaded? _____%

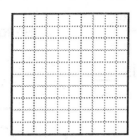

One half of something is _____% of it.

$$\frac{1}{2} = \underline{\hspace{2cm}}\%$$

Put hair on half the heads.
Put ears on 50% of the heads.

Circle 50% of the dollar.

Put **X**'s in 50% of the squares.
Shade 50% of the squares.

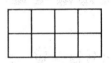

Shade 50% of the shapes.

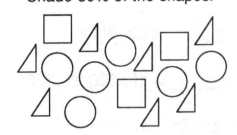

Circle half of the stars.

Shade about 50% of the circle.

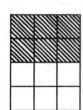 How many squares? **12**
Shade 50%.
How many shaded? **6**

50% of __**12**__ is __**6**__.

| How many squares? ____
Shade 50%.
How many shaded? ____

50% of _____ is _____.

 How many flowers? ____
Circle 50%.
How many circled? ____

$\frac{1}{2}$ of _____ is _____.
50% of _____ is _____.

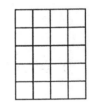 How many bulbs? ____
Circle 50%.
How many circled? ____

$\frac{1}{2}$ of _____ is _____.
50% of _____ is _____.

Shade **one fourth** of the square.

What percent of the
square is shaded? _____%

One fourth of something is _____% of it.

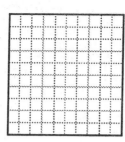

$\dfrac{1}{4}$ = _____%

Put hair on 25% of the heads.
Put ears on 100% of the heads.

Circle 25% of the stars.

Shade about 25% of the circle.

How many squares? __12__
Shade 25%.
How many shaded? ____

25% of _____ is _____.

How many squares? ____
Shade 25%.
How many shaded? ____

25% of _____ is _____.

How many stars? ____
Circle 25%.
How many circled? ____

$\dfrac{1}{4}$ of _____ is _____.
25% of _____ is _____.

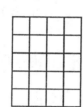

How many circles? ____
Shade 25%.
How many shaded? ____

$\dfrac{1}{4}$ of _____ is _____.
25% of _____ is _____.

Try these without pictures to help.

$\dfrac{1}{2}$ of 28 is __14__.   50% of 28 is __14__. | $\dfrac{1}{4}$ of 28 is ____.   25% of 28 is ____.

$\dfrac{1}{2}$ of 36 is ____.   ____% of 36 is 18. | $\dfrac{1}{4}$ of 36 is ____.   ____% of 36 is 9.

Since 50% = $\frac{1}{2}$ and 25% = $\frac{1}{4}$, there is an easy way to find 50% and 25% of a number.

> To find **50%** of a number, simply **divide by 2**.
> To find **25%** of a number, simply **divide by 4**.

Use division to fill in the chart below.

100%	32	80	200	24¢	$1.80	$12.40	$104
50%	16						
25%	8						

Use division to solve each problem below.

50% of 40 is _____. {40÷2}

25% of 40 is _____. {40÷4}

50% of $44 is _____.

25% of $44 is _____.

50% of $9.00 is _____.

25% of $9.00 is _____.

6 is 50% of _____.

6 is _____% of 12.

_____ is 50% of 12.

_____ is 25% of 48.

12 is _____% of 48.

12 is 25% of _____.

_____ is 25% of $20.

$5 is _____% of $20.

$5 is 25% of _____.

9 is _____% of 18.

2 is _____% of 8.

$1 is _____% of $2.

_____ is 50% of 30.

_____ is 25% of 36.

_____ is 25% of $64.

3 is 50% of _____.

10 is 50% of _____.

$7 is 25% of _____.

Be careful on the ones below! They're trickier.

$1.50 is _____% of $3.

$14.25 is 50% of _____.

_____ is 25% of $8.00.

_____ is 25% of $5.00.

$3.50 is 25% of _____.

_____ is 25% of $42.00.

100% Makes It All

Fill in each missing percent. Remember, all of something is _____% of it.

42%
58%
25%
50%
25%
35%
20%
10%
30%
35%
25%
22%
19%
25%
8%
32%
46%

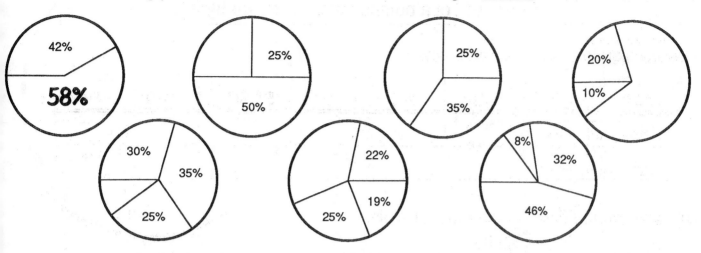

A **pie chart** is one useful way to show information. A pie chart is made with a circle divided into slices of "pie." The complete circle is 100%.

Fill in the missing percent in each pie chart below. Then use the pie chart to help answer the question.

Students at Cabot High School

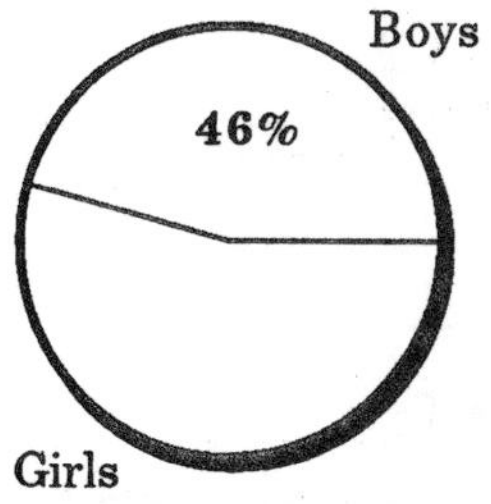

Boys
46%
Girls

What percent of the students at Cabot High are girls?

____

Favorite Pet of 7th Graders

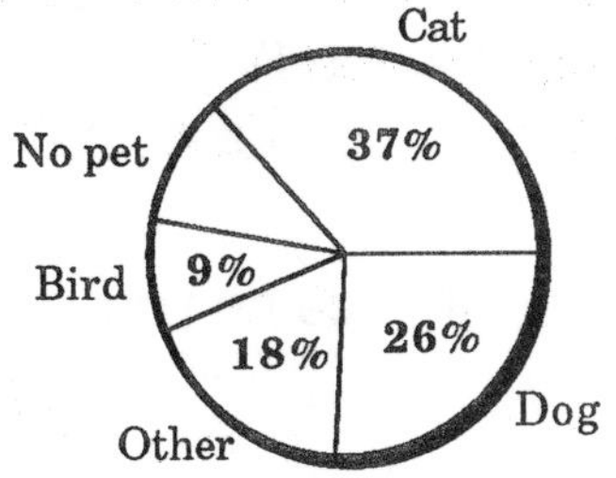

Cat
No pet
37%
Bird
9%
18%
26%
Dog
Other

What percent of the seventh graders chose no pet?

____

Motor Vehicles at the Stadium Lot

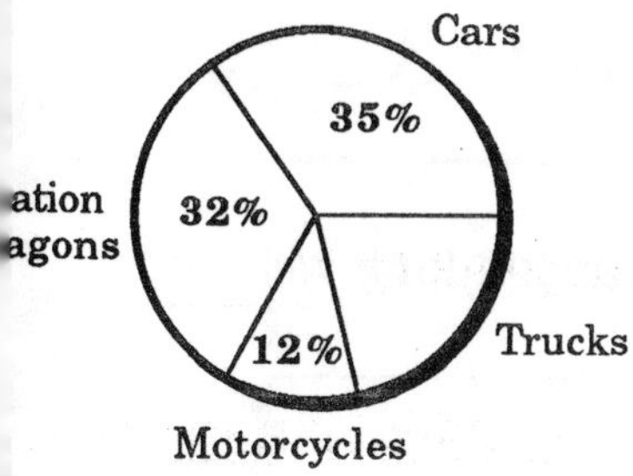

Cars
35%
ation
agons
32%
12%
Trucks
Motorcycles

Trucks and cars make up what percent of the vehicles?

____

Math Test Results

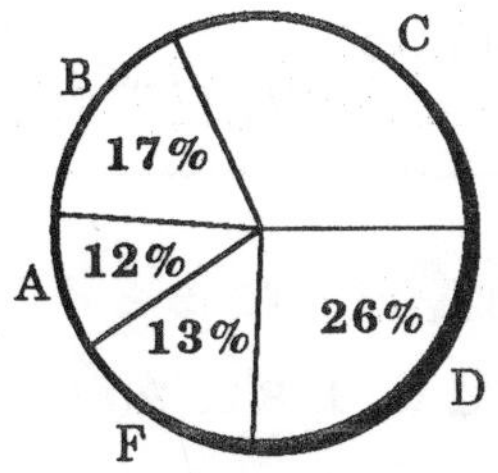

B
C
17%
A
12%
13%
26%
D
F

What percent of the students got higher than a D on the test?

____

The diagram shows how body weight is divided between the body parts in the average adult in the U.S. Each arm is about the same weight. What percent of a body's weight is in each arm?

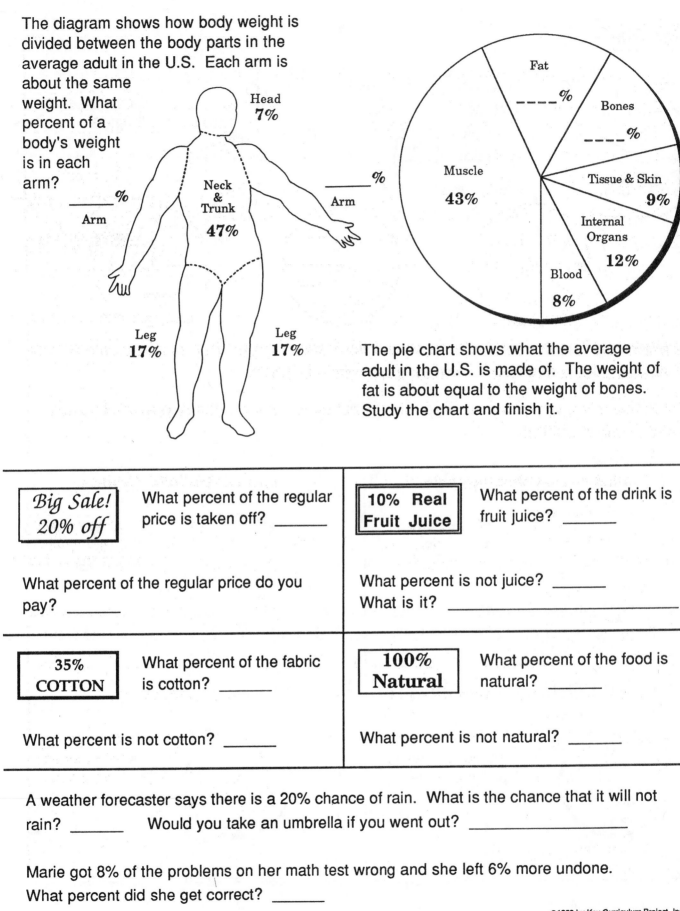

Head
7%

Neck & Trunk
47%

_____ %
Arm

_____ %
Arm

Leg
17%

Leg
17%

Fat
----- %

Bones
----- %

Muscle
43%

Tissue & Skin
9%

Internal Organs
12%

Blood
8%

The pie chart shows what the average adult in the U.S. is made of. The weight of fat is about equal to the weight of bones. Study the chart and finish it.

| Big Sale! 20% off | What percent of the regular price is taken off? _____ |

What percent of the regular price do you pay? _____

| 10% Real Fruit Juice | What percent of the drink is fruit juice? _____ |

What percent is not juice? _____
What is it? _____

| 35% COTTON | What percent of the fabric is cotton? _____ |

What percent is not cotton? _____

| 100% Natural | What percent of the food is natural? _____ |

What percent is not natural? _____

A weather forecaster says there is a 20% chance of rain. What is the chance that it will not rain? _____   Would you take an umbrella if you went out? _____

Marie got 8% of the problems on her math test wrong and she left 6% more undone. What percent did she get correct? _____

Floor Plans

Ramona Rodriguez is an architect. She has designed 100 square meters of new office space for a high school. Look at her plan. Use it to answer each question below.

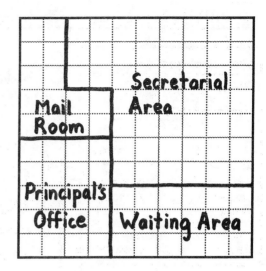

The mail room is _____% of the area.

The principal's office takes up _____% of the space.

The secretarial area occupies _____% of the space.

_____% of the office is taken up by the waiting area.

Now it's your turn to be the architect. You have been hired by a school to design a new art studio. The studio will be 100 square meters. Draw walls and label the areas using the guidelines below.

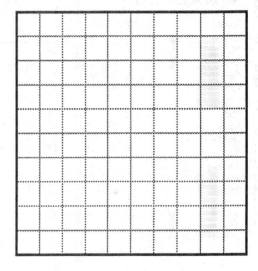

6% of the area is the teacher's office.

6% is a storage room.

21% is a ceramics area.

9% is a kiln room.

The rest is the classroom.
What percent is classroom? _____%

You did so well designing the school building that you've been hired by the principal to plan a new summer cottage for his family. The cottage will be 100 square meters (including the porch). Make a floor plan for the cottage.

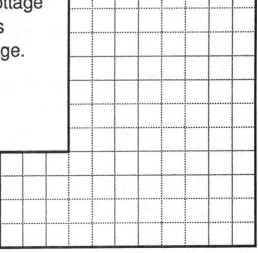

22% of the area is living room.

Two bedrooms each occupy 15%.

20% is kitchen.

6% is bathroom.

12% is porch.

The rest is hall and storage area.
What percent is hall and storage? _____%

Recommended Daily Allowances (RDA) of Vitamins

Nutrition Information Per Serving

Serving Size: 1 Oz.
(About ¼ Cup) (28.35 g)
Servings Per Package: 24

	1 OZ. (28.35 g) Cereal	With ½ Cup (118 mL) Vitamin D Fortified Whole Milk
Calories	110	180*
Protein	3 g	7 g
Carbohydrate	23 g	29 g
Fat	0	4 g
Sodium	190 mg	250 mg

Percentages of U.S. Recommended Daily Allowances (U.S. RDA)

Protein	4%	10%
Vitamin A	25%	30%
Vitamin C	**	**
Thiamine	25%	30%
Riboflavin	25%	35%
Niacin	25%	25%
Calcium	**	15%
Iron	15%	15%
Vitamin D	10%	25%
Vitamin B6	25%	30%
Folic Acid	25%	25%
Vitamin B12	25%	30%
Phosphorus	6%	15%
Magnesium	6%	10%
Zinc	8%	10%
Copper	6%	6%

*Save 30 calories – use skim milk.
**Contains less than 2% of the U.S. RDA of these nutrients.

This label is from a box of breakfast cereal. It tells about the amounts of vitamins and minerals in a one ounce serving of cereal. The amounts are given as percentages of the Recommended Daily Allowance (RDA) of each vitamin and mineral.

The left column gives the percentages in one ounce of cereal alone. The right column gives the percentages in one ounce of cereal with one half cup of milk.

Use the information to complete the tables below.

Nutrient	1 ounce serving of cereal	Cereal with half cup milk
Protein	4%	10%
Vitamin A		
Niacin		

Nutrient	1 ounce serving of cereal	Cereal with half cup milk
Vitamin D		
Zinc		
Copper		

10% – 4%

What percent of the RDA of protein do you get in $\frac{1}{2}$ cup of milk alone? _____

What percent of the RDA of vitamin D do you get in $\frac{1}{2}$ cup of milk alone? _____

What percent of the RDA of copper do you get in $\frac{1}{2}$ cup of milk alone? _____

What percent of the RDA of protein do you still need after you've eaten one serving of cereal alone? _____

What percent of the RDA of vitamin A do you still need after you've eaten one serving of cereal with milk? _____

How many servings of cereal would you need to eat to get 100% of the RDA of niacin? _____

14

Double and Triple as Percents

Shade **100%** of a square.

Shade another **100%** of a square.

All together you shaded _____% of a square.

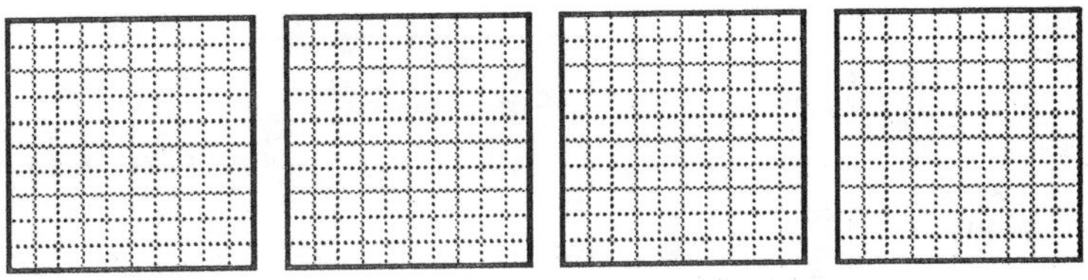

All together you shaded ___ squares.

Two of something is _____% of it.

2 = _____%

Shade **100%** of a square.

Shade a second **100%** of a square.

Shade a third **100%** of a square.

All together you shaded _____% of a square.

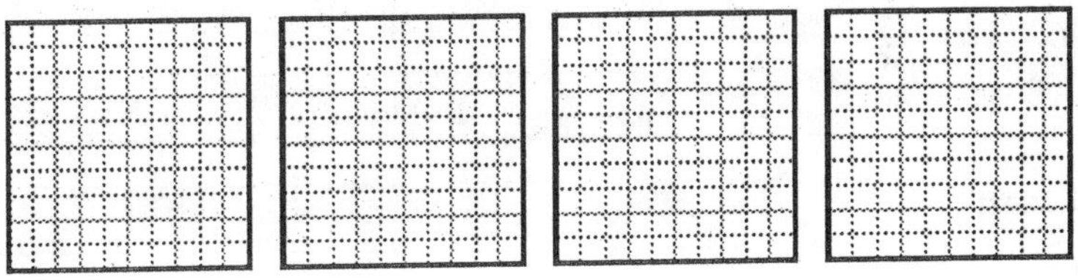

All together you shaded ___ squares.

Three of something is _____% of it.

3 = _____%

Shade 200% of a triangle.

Shade 300% of a circle.

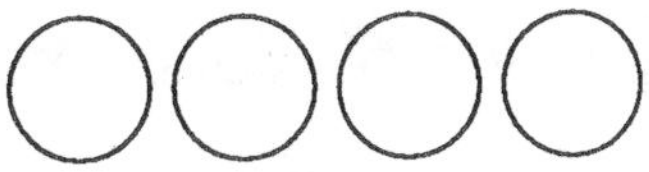

Shade 400% of a square.

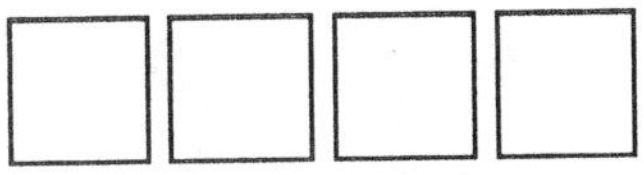

100% of 6 is _____.

200% of 6 is _____.

300% of 6 is _____.

32 is _____% of 16.

100% of 15 is _____.

200% of 15 is _____.

300% of 15 is _____.

50 is _____% of 25.

_____ is 200% of 8.

_____ is 300% of 8.

_____ is 400% of 8.

27 is _____% of 9.

Below is a graph of the average annual precipitation (rain and melted snow) for 10 cities in North America. The actual precipitation is given at the end of each bar in the graph.

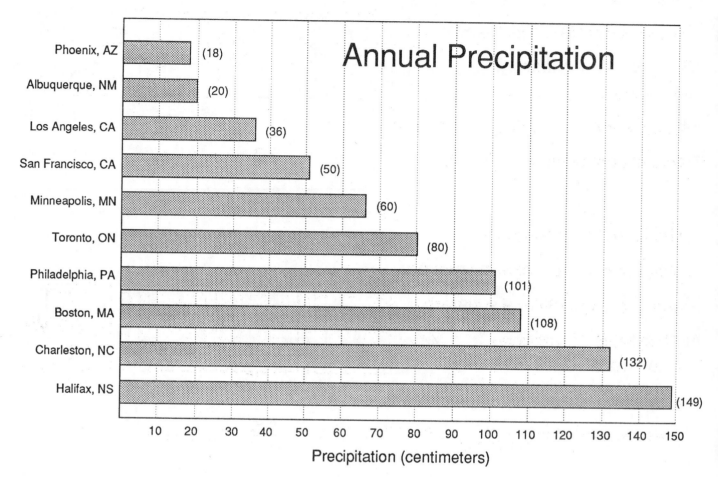

Annual Precipitation

Phoenix, AZ (18)
Albuquerque, NM (20)
Los Angeles, CA (36)
San Francisco, CA (50)
Minneapolis, MN (60)
Toronto, ON (80)
Philadelphia, PA (101)
Boston, MA (108)
Charleston, NC (132)
Halifax, NS (149)

Precipitation (centimeters)

Which city has 200% of the precipitation of Phoenix? _____

Which city has 300% of the precipitation of Los Angeles? _____

Which city has 400% of the precipitation of Albuquerque? _____

Which city has 600% of the precipitation of Phoenix? _____

Which city has 250% of the precipitation of Albuquerque? _____

Which city has about 300% of the precipitation of San Francisco? _____

Which city has about 400% of the precipitation of Los Angeles? _____

Which city has about 500% of the precipitation of Albuquerque? _____

Which city has about 250% of the precipitation of Minneapolis? _____

Which city has about 750% of the precipitation of Albuquerque? _____

16

Practice with Percents

Draw hats on 50% of the heads.

Put smiles on 100% of the heads.

Give 25% of the heads ears.

Paint 0% of the heads blue.

Shade 25% of all the rectangles.

Put an **X** inside 50%.

Make a ring around 100%.

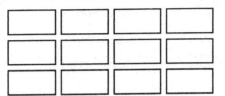

Show how much each person earns per hour.

Sally earns $6.00 per hour.

Sally: | $5 | | $1 | $ _6.00_

Ted earns 200% of Sally's pay.

Ted: $_____

Ora earns 300% of Sally's pay.

Ora: $_____

Alisa earns 50% of Sally's pay.

Alisa: $_____

Joan earns 200% of Ted's pay.

Joan: $_____

How full should each cup be? Shade the cups to show your answer.

Cup A is 25% full.

Cup B has 200% as much as cup A.

Cup C has 50% as much as cup B.

Cup D has 300% as much as cup A.

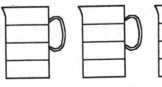

A   **B**   **C**   **D**

What percent of the things in each group are circled? What percent are not circled?

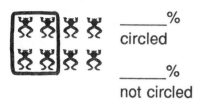 _____% circled

_____% not circled

 _____% circled

_____% not circled

 _____% circled

_____% not circled

 _____% circled

_____% not circled

 _____% circled

_____% not circled

 _____% circled

_____% not circled

What percent of each square is shaded?

_____% is shaded.

_____% is shaded.

_____% is shaded.

_____% is shaded.

_____% is shaded.

Answer each question using a complete percent statement.

Problem:	What is 50% of $24?	Problem:	What is 100% of $17?
Statement:	**$12 is 50% of $24.**	Statement:	
Problem:	What is 200% of $12?	Problem:	What is 50% of $5?
Statement:		Statement:	
Problem:	What is 25% of $12?	Problem:	What is 50% of $2.50?
Statement:		Statement:	
Problem:	What is 200% of $6.75?	Problem:	What is 25% of $5?
Statement:		Statement:	

Draw each line segment.

$\overline{AB}$

50% as long as $\overline{AB}$

25% as long as $\overline{AB}$

$\overline{CD}$

25% as long as $\overline{CD}$

100% as long as $\overline{CD}$

$\overline{EF}$

100% as long as $\overline{EF}$

200% as long as $\overline{EF}$

$\overline{GH}$

50% as long as $\overline{GH}$

200% as long as $\overline{GH}$

$\overline{IJ}$

200% as long as $\overline{IJ}$

300% as long as $\overline{IJ}$

$\overline{KL}$

50% as long as $\overline{KL}$

100% as long as $\overline{KL}$

150% as long as $\overline{KL}$

0% as long as $\overline{KL}$

$\overline{MN}$

25% as long as $\overline{MN}$

50% as long as $\overline{MN}$

75% as long as $\overline{MN}$

100% as long as $\overline{MN}$

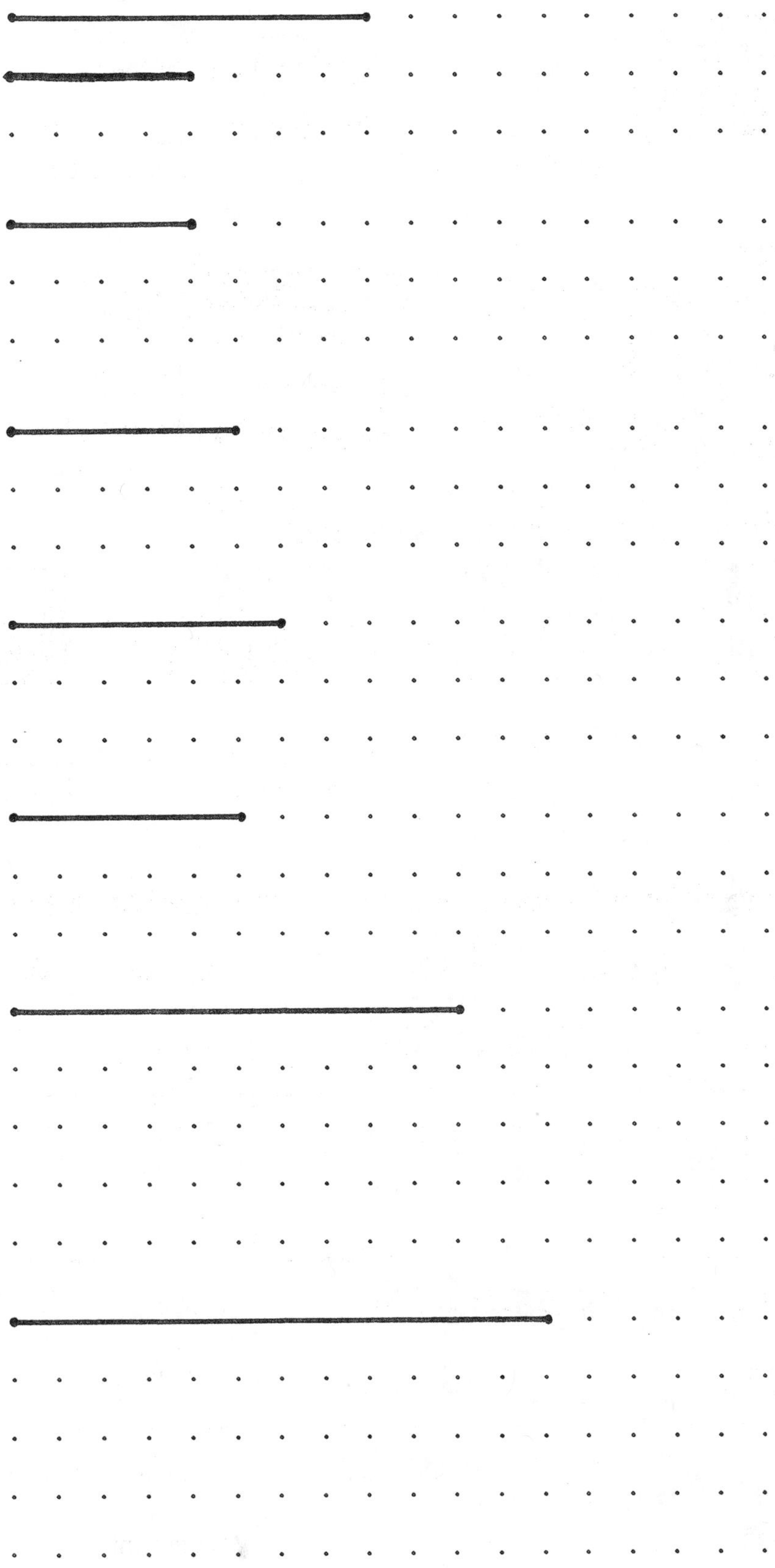

0% of 60 is _____.   25% of 44 is _____.   0% of $21 is _____.

25% of 60 is _____.   50% of 44 is _____.   100% of $21 is _____.

50% of 60 is _____.   75% of 44 is _____.   200% of $21 is _____.

100% of 60 is _____.   100% of 44 is _____.   300% of $21 is _____.

_____ is 0% of 320.   _____ is 25% of 140.   _____ is 0% of $85.

_____ is 25% of 320.   _____ is 50% of 140.   _____ is 100% of $85.

_____ is 50% of 320.   _____ is 75% of 140.   _____ is 200% of $85.

_____ is 100% of 320.   _____ is 100% of 140.   _____ is 300% of $85.

300% of $5 is _____.   300% of $27 is _____.

_____ is 0% of $200.   _____ is 50% of $220.

100% of 36492 is _____.   0% of 43621 is _____.

50% of 222222 is _____.   100% of 9312 is _____.

0% of 1123456789 is _____.   300% of 11111 is _____.

100% of $6.35 is _____.   0% of $1835.25 is _____.

50% of $25 is $12.50 _____. ⟨½ of $25⟩   _____ is 50% of $21.

50% of $19 is _____. ⟨½ of $19⟩   _____ is 50% of $7.

25% of $30 is _____. ⟨¼ of $30⟩   _____ is 25% of $50.

20

One Tenth as a Percent

Shade ten small squares.

What percent of the
large square is shaded? _____%

You shaded $\frac{10}{100}$ or $\frac{1}{10}$ of the large square.

One tenth of something is _____% of it.

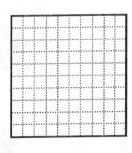

$\frac{1}{10}$ = _____%

To find **10%** of a number, simply **divide by 10**.

{50÷10}

$\frac{1}{10}$ of 50 is _____.   10% of 50 is _____.

{20÷10}

$\frac{1}{10}$ of 20 is _____.   10% of 20 is _____.

$\frac{1}{10}$ of 900 is _____.   10% of 900 is _____.

$\frac{1}{10}$ of 40 is _____.   10% of 40 is _____.

Dividing by 10 is easy when the number you're dividing ends in zero. You don't need
pencil and paper. Simply drop the last zero.

30
100% is _____.
10% is _____.

70
100% is _____.
10% is _____.

120
100% is _____.
10% is _____.

300
100% is _____.
10% is _____.

10% of 10 is __*l*__ .

10% of 20 is _____ .

10% of 30 is _____ .

10% of 40 is _____ .

_____ is 10% of 80.

5 is 10% of _____ .

10% of 60 is _____ .

10% of 120 is _____ .

10% of 180 is _____ .

10% of 240 is _____ .

_____ is 10% of 390.

48 is 10% of _____ .

10% of $750 is _____ .

10% of $1500 is _____ .

10% of $2250 is _____ .

10% of $3000 is _____ .

_____ is 10% of 4560.

125 is 10% of _____ .

Each pie chart has been divided into equal parts. What percent is shaded? What percent is not shaded?

 _____% shaded

_____% not shaded

 _____% shaded

_____% not shaded

 _____% shaded

_____% not shaded

 _____% shaded

_____% not shaded

 _____% shaded

_____% not shaded

 _____% shaded

_____% not shaded

If you know 10% of a number, you can easily find 20%, 30%, or 40% of the number.

10% of 90 is __9__.

20% of 90 is __18__. (2 x 9)

30% of 90 is _____. (3 x ___)

40% of 90 is _____. (___ x ___)

50% of 90 is _____.

60% of 90 is _____.

70% of 90 is _____.

80% of 90 is _____.

90% of 90 is _____.

100% of 90 is _____.

110% of 90 is _____.

120% of 90 is _____.

10% of 300 is _____.

20% of 300 is _____.

30% of 300 is _____.

40% of 300 is _____.

50% of 300 is _____.

60% of 300 is _____.

70% of 300 is _____.

80% of 300 is _____.

90% of 300 is _____.

100% of 300 is _____.

110% of 300 is _____.

120% of 300 is _____.

_____ is 10% of $50.

_____ is 20% of $50.

_____ is 30% of $50.

10% of 240 is _____.

20% of 240 is _____.

30% of 240 is _____.

10% of 400 is _____.

20% of 400 is _____.

30% of 400 is _____.

One Hundredth as a Percent

Shade one small square.

What percent of the
large square is shaded? _____%

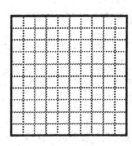

You shaded $\frac{1}{100}$ of the large square.

One hundredth of something is _____% of it. $\frac{1}{100}$ = _____%

To find **1%** of a number, simply **divide by 100**.

$300 \div 100$

$\frac{1}{100}$ of 300 is ____.   1% of 300 is ___.

$500 \div 100$

$\frac{1}{100}$ of 500 is ____.   1% of 500 is ___.

$\frac{1}{100}$ of 700 is ____.   1% of 700 is ___. $\frac{1}{100}$ of 900 is ____.   1% of 900 is ___.

Dividing by 100 is easy when the number you're dividing ends in two zeros. Simply drop the last two zeros.

300	700	1200	3000
100% is _____.	100% is _____.	100% is _____.	100% is _____.
1% is _____.	1% is _____.	1% is _____.	1% is _____.

1% of 100 is _____.   1% of 400 is _____.   1% of $2300 is _____.

1% of 200 is _____.   1% of 800 is _____.   1% of $4600 is _____.

1% of 300 is _____.   1% of 1200 is _____.   1% of $6900 is _____.

1% of 400 is _____.   1% of 1600 is _____.   1% of $9200 is _____.

_____ is 1% of 900.   _____ is 1% of 3900.   _____ is 1% of 9900.

5 is 1% of _____.   48 is 1% of _____.   125 is 1% of _____.

If you know 1% of a number, you can easily find 2%, 3%, or 4% of the number.

1% of 600 is ___6___.   1% of 1200 is _____.

2% of 600 is _____. {2 x 6}   2% of 1200 is _____.

3% of 600 is _____. {__ x 6}   3% of 1200 is _____.

4% of 600 is _____.   4% of 1200 is _____.

5% of 600 is _____.   5% of 1200 is _____.

6% of 600 is _____.   6% of 1200 is _____.

_____ is 1% of 900.   1% of 1500 is _____.   1% of $7500 is _____.

_____ is 2% of 900.   2% of 1500 is _____.   2% of $7500 is _____.

_____ is 3% of 900.   3% of 1500 is _____.   3% of $7500 is _____.

_____ is 4% of 900.   4% of 1500 is _____.   4% of $7500 is _____.

_____ is 5% of 900.   5% of 1500 is _____.   5% of $7500 is _____.

_____ is 6% of 900.   6% of 1500 is _____.   6% of $7500 is _____.

_____ is 7% of 900.   7% of 1500 is _____.   7% of $7500 is _____.

_____ is 8% of 900.   8% of 1500 is _____.   8% of $7500 is _____.

_____ is 9% of 900.   9% of 1500 is _____.   9% of $7500 is _____.

_____ is 10% of 900.   10% of 1500 is _____.   10% of $7500 is _____.

Each group of problems has a pattern. Finding the pattern can help you get the answer quickly.

100%	200		
1%	2	5	
2%	4		
4%			100
8%			
16%			

100%	600		
1%		8	
3%			
9%			
27%			
81%			162

100%			1100
1%	3		
5%			
10%		70	
50%			

Practice with 1% and 10%

600
100% is _____.
10% is _____.
1% is _____.

400
100% is _____.
10% is _____.
1% is _____.

1700
100% is _____.
10% is _____.
1% is _____.

7000
100% is _____.
10% is _____.
1% is _____.

900
100% is _____.
_____% is 90.
1% is _____.

500
_____% is 500.
10% is _____.
1% is _____.

_____
100% is 4100.
_____% is 410.
_____% is 41.

_____
100% is _____.
10% is 500.
1% is _____.

100% of 200 is _____.
10% of 200 is _____.
1% of 200 is _____.

100% of 1300 is _____.
10% of 1300 is _____.
1% of 1300 is _____.

100% of $5400 is _____.
10% of $5400 is _____.
1% of $5400 is _____.

_____ is 10% of 60.
_____ is 1% of 600.

_____ is 10% of 190.
_____ is 1% of 1900.

_____ is 10% of $730.
_____ is 1% of $7300.

4 is _____% of 4.
4 is _____% of 40.
4 is _____% of 400.

2 is _____% of 200.
2 is _____% of 2.
2 is _____% of 20.

$50 is _____% of $5000.
$50 is _____% of $500.
$50 is _____% of $50.

Make up some problem sets of your own.

1% of _____ is _____.
10% of _____ is _____.
100% of _____ is _____.

1% of _____ is _____.
10% of _____ is _____.
100% of _____ is _____.

1% of _____ is _____.
10% of _____ is _____.
100% of _____ is _____.

100¢ = $1.00   One hundred cents equals one dollar.

10% of $1.00 = $.10   1% of $1.00 = $.01

Finding 10% or 1% of dollar amounts can be confusing when the amount is expressed as a decimal. It helps to think of the amount as all cents.

True or false?

10% of $.90 is $.09
(True)   False

10% of $3.00 is $.03
True   False

10% of $.50 is $.05
True   False

10% of $1.50 is $1.05
True   False

10% of $2.00 is $.20
True   False

10% of $14.50 is $1.45
True   False

10% of $.70 is $7.00
True   False

10% of $16.00 is $1.60
True   False

10% of $.29 is $.29
True   False

Remember, in order to properly show cents with a decimal point, the decimal must show hundredths.

Doesn't show hundredths.

10% of $1.60 is $.160
True   (False)

10% of $4.70 is $.047
True   False

10% of $300 is $30
True   False

1% of $600 is $6
True   False

1% of $7.00 is $.07
True   False

1% of $800 is $8
True   False

1% of $300.00 is $30.00
True   False

1% of $12 is $.12
True   False

1% of $53 is $5.30
True   False

26

Complete each sales receipt.

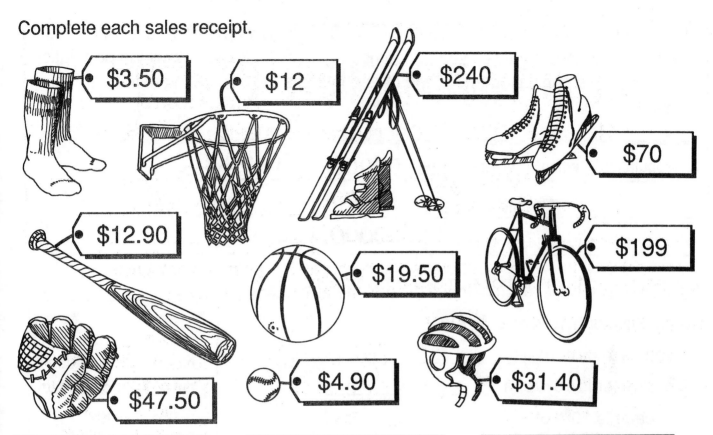

Buena Sports		
Item	Price	
1 Pair Skates	70	00
1 Pair Socks	3	50
Subtotal	73	50
10% Tax	7	35
Total	80	85

Buena Sports		
Item	Price	
1 Basketball		
1 Hoop		
Subtotal		
10% Tax		
Total		

Buena Sports		
Item	Price	
1 Ski Set		
1 Pair Skates		
Subtotal		
10% Tax		
Total		

Buena Sports		
Item	Price	
1 Bicycle		
1 Helmet		
Subtotal		
10% Tax		
Total		

Buena Sports		
Item	Price	
1 Bat		
1 Softball		
1 Glove		
Subtotal		
10% Tax		
Total		

Make up your own.

Buena Sports		
Item	Price	
Subtotal		
10% Tax		
Total		

The population of Rodeo City is 6,000,000. City planners expect it to grow by 1% each year for the next three years. Complete the chart to show how the population will grow. The population at the end of each year is the population at the beginning of the next year.

	Year 1	Year 2	Year 3
Population (Beginning of year)	6000000	6060000	
1% Increase	60000		
Population (End of year)	6060000		

Choose a word from the box to make each sentence true.

A cent is worth 10% of a __dime_____.

A year is 10% of a _____.

A millimeter is 10% of a _____.

A decade is 10% of a _____.

A dime is 10% of a _____.

month	year
nickel	decade
~~dime~~	centimeter
meter	penny
century	dollar

Make three different true sentences using words from the box.

A _____ is 1% of a _____ .

A _____ is 1% of a _____ .

A _____ is 1% of a _____ .

penny	century
centimeter	dollar
meter	year

Fill in each blank.

10% of 1 minute is _____ seconds.

10% of 2 minutes is _____ seconds.

10% of 3 minutes is _____ seconds.

10% of 1 hour is _____ minutes.

10% of 2 hours is _____ minutes.

10% of 3 hours is _____ minutes.

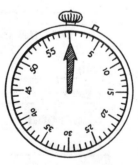

1 day = 24 hours
1 hour = 60 minutes

10% of 1 day is _____ minutes.

Other Percents

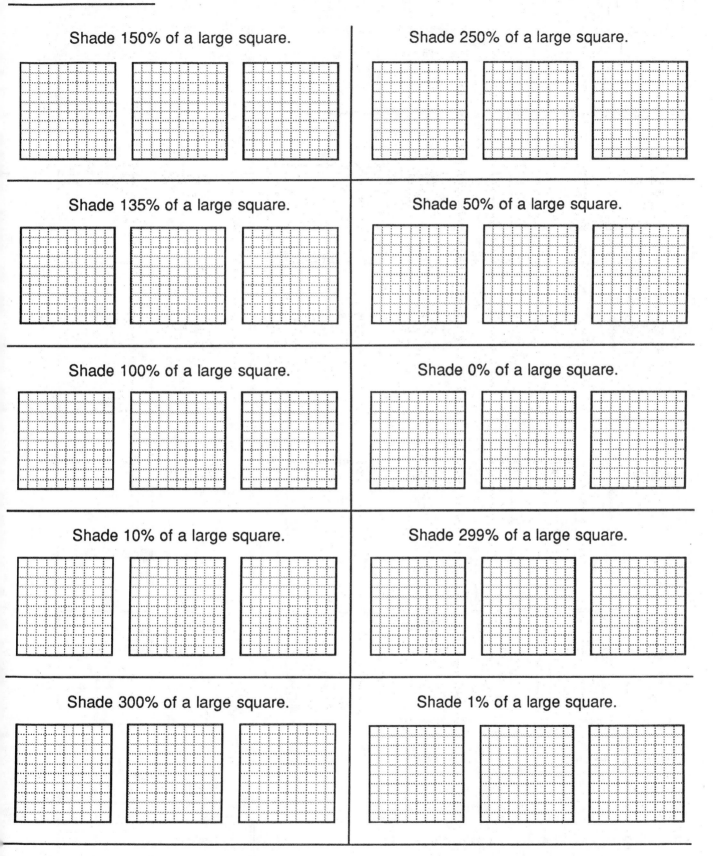

Shade 150% of a large square.

Shade 250% of a large square.

Shade 135% of a large square.

Shade 50% of a large square.

Shade 100% of a large square.

Shade 0% of a large square.

Shade 10% of a large square.

Shade 299% of a large square.

Shade 300% of a large square.

Shade 1% of a large square.

Work your way down each column. Can you find the pattern?

Shade
16%.
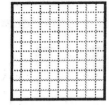

Shade
64%.
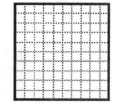

Shade
81%.
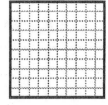

Shade
8%.
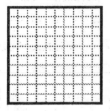

Shade
16%.
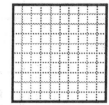

Shade
27%.
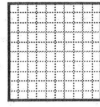

Shade
4%.
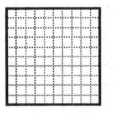

Shade
4%.
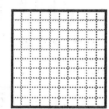

Shade
9%.
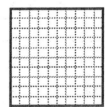

Shade
2%.
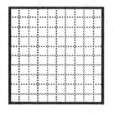

Shade
1%.
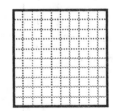

Shade
3%.
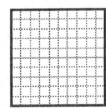

Shade
1%.
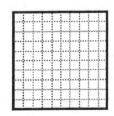

Shade
$\frac{1}{4}$%.
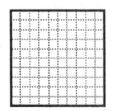

Shade
1%.
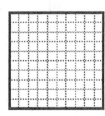

Shade
$\frac{1}{2}$%.
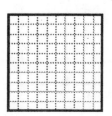

Shade
$\frac{1}{3}$%.
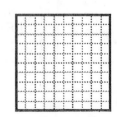

100¢ = $1.00   One hundred cents equals one dollar.

_____% of a dollar

_____% of a dollar

_____% of $1.00

_____% of $1.00

Use percent notation or a dollar amount to make each statement true.

$.50 is _____ of a dollar.

$.25 is _____ of a dollar.

$.75 is _____ of a dollar.

_____ is 98% of a dollar.

_____ is 28% of $1.00.

$1.60 is _____ of $1.00.

_____ is 46% of $1.00.

_____ is 100% of $1.00.

$2.50 is _____ of $1.00.

$9.38 is _____ of $1.00.

$1.00 is _____ of $2.00.

$.50 is _____ of $2.00.

$.20 is _____ of $2.00.

$.02 is _____ of $2.00.

A dime is _____ of a dollar.

Three quarters is _____ of a dollar.

A nickel plus a dime plus a quarter is _____ of a dollar.

Make up your own!

_____ is _____ of one dollar.

_____ is _____ of two dollars.

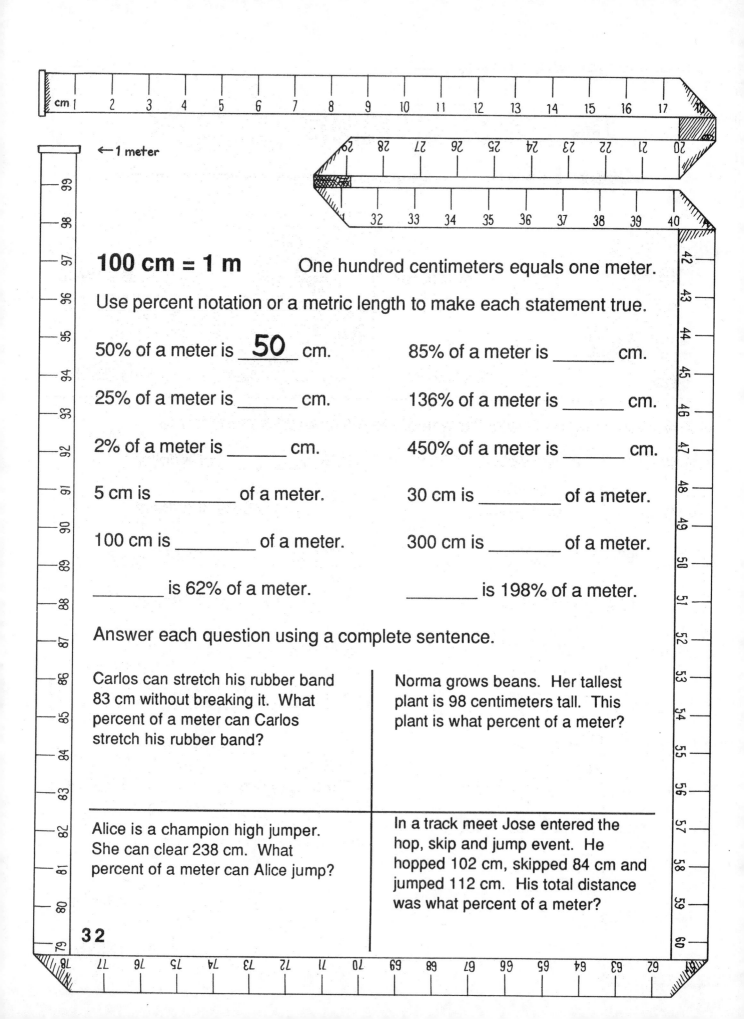

100 cm = 1 m   One hundred centimeters equals one meter.

Use percent notation or a metric length to make each statement true.

50% of a meter is __**50**__ cm.   85% of a meter is _____ cm.

25% of a meter is _____ cm.   136% of a meter is _____ cm.

2% of a meter is _____ cm.   450% of a meter is _____ cm.

5 cm is _____ of a meter.   30 cm is _____ of a meter.

100 cm is _____ of a meter.   300 cm is _____ of a meter.

_____ is 62% of a meter.   _____ is 198% of a meter.

Answer each question using a complete sentence.

Carlos can stretch his rubber band 83 cm without breaking it. What percent of a meter can Carlos stretch his rubber band?

Norma grows beans. Her tallest plant is 98 centimeters tall. This plant is what percent of a meter?

Alice is a champion high jumper. She can clear 238 cm. What percent of a meter can Alice jump?

In a track meet Jose entered the hop, skip and jump event. He hopped 102 cm, skipped 84 cm and jumped 112 cm. His total distance was what percent of a meter?

32

Finding a Percent of a Number

Match.

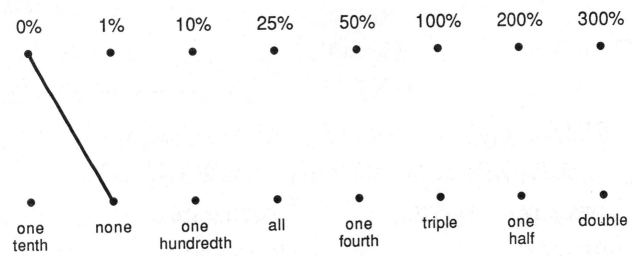

| 0% | 1% | 10% | 25% | 50% | 100% | 200% | 300% |

| one tenth | none | one hundredth | all | one fourth | triple | one half | double |

Choose the correct ending for each sentence from the box on the right.

To find 0% of a number you **write zero** _____.

To find 1% of a number you _____.

To find 10% of a number you _____.

To find 25% of a number you _____.

To find 50% of a number you _____.

To find 100% of a number you _____.

To find 200% of a number you _____.

To find 300% of a number you _____.

> divide by 4
>
> multiply by 2
>
> write the number
>
> ~~write zero~~
>
> divide by 100
>
> divide by 10
>
> multiply by 3
>
> divide by 2

Make each statement true.

25% of 400 is **400** divided by **4**.   0% of 400 is _____.

10% of 400 is _____ divided by _____.   1% of 400 is _____ divided by _____.

100% of 400 is _____.   300% of 400 is _____ times _____.

50% of 400 is _____ _____ _____.

200% of 400 is _____ _____ _____.

25% of 16 is _____.   25% of $80 is _____.   _____ is 25% of 400.

50% of 16 is _____.   50% of $80 is _____.   _____ is 50% of 400.

75% of 16 is _____.   75% of $80 is _____.   _____ is 75% of 400.

100% of 16 is _____.   100% of $80 is _____.   _____ is 100% of 400.

125% of 16 is _____.   125% of $80 is _____.   _____ is 125% of 400.

           0% of 60 is _____.   0% of 120 is _____.

Start   10% of 60 is _____.   Start   10% of 120 is _____.
here.   here.

        25% of 60 is _____.   25% of 120 is _____.

    50% of 60 is _____.   50% of 120 is _____.

100% of 60 is _____.   100% of 120 is _____.

   200% of 60 is _____.   200% of 120 is _____.

    300% of 60 is _____.   300% of 120 is _____.

        1% of $200 is _____.   1% of 3700 is _____.

     10% of $200 is _____.   10% of 3700 is _____.

    25% of $200 is _____.   25% of 3700 is _____.

   50% of $200 is _____.   50% of 3700 is _____.

100% of $200 is _____.   100% of 3700 is _____.

  200% of $200 is _____.   200% of 3700 is _____.

   300% of $200 is _____.   300% of 3700 is _____.

100% of 48 is _____.   10% of 120 is _____.

50% of 48 is _____.   5% of 120 is _____. ⎨ ½ of 10% ⎬

25% of 48 is _____.   2.5% of 120 is _____. ⎨ ½ of 5% ⎬

12.5% of 48 is _____. ⎨ ½ of 25% ⎬   1.25% of 120 is _____.

5 is _____ of 500.

50 is _____ of 500.

125 is _____ of 500.

250 is _____ of 500.

500 is **100%** of 500.

1000 is _____ of 500.

1500 is _____ of 500.

$4 is _____ of $400.

$40 is _____ of $400.

$100 is _____ of $400.

$200 is _____ of $400.

$400 is _____ of $400.

$800 is _____ of $400.

$1200 is _____ of $400.

0 is _____ of 120.

30 is _____ of 120.

60 is _____ of 120.

90 is _____ of 120.

120 is _____ of 120.

1000 is _____ of 1000.

100 is _____ of 1000.

200 is _____ of 1000.

300 is _____ of 1000.

400 is _____ of 1000.

$96 is _____ of $96.

$48 is _____ of $96.

$24 is _____ of $96.

$72 is _____ of $96.

$0 is _____ of $96.

$15 is _____ of $150.

$30 is _____ of $150.

$45 is _____ of $150.

$60 is _____ of $150.

$75 is _____ of $150.

$90 is _____ of $150.

$105 is _____ of $150.

$120 is _____ of $150.

$135 is _____ of $150.

$150 is _____ of $150.

$165 is _____ of $150.

$7 is _____ of $70.

$14 is _____ of $70.

$21 is _____ of $70.

$28 is _____ of $70.

$35 is _____ of $70.

$5 is _____ of $50.

$10 is _____ of $50.

$20 is _____ of $50.

$40 is _____ of $50.

$80 is _____ of $50.

What percent of the problems on this page do you think you did correctly? _____

Pat took the test below. Put C by each problem Pat did correctly and X by each answer that is wrong. Don't write anything by the problems that Pat did not do.

	Pat
	Percent Quiz
○	1. 100% of 53 is <u>53</u>. **C** 11. 10% of 1500 is ___.
	2. 0% of 53 is <u>53</u>. **X** 12. 1% of 200 is ___.
	3. 50% of 128 is <u>68</u>. 13. 1% of 1500 is ___.
	4. 50% of $20 is <u>$10</u>. 14. 1% of 300 is <u>3</u>.
	5. 50% of 48cm is <u>24cm</u>. 15. 2% of 300 is <u>6</u>.
	6. 50% of 21 is ___. 16. 3% of 300 is ___.
	7. 25% of 32 is <u>8</u>. 17. 6% of 300 is <u>12</u>.
○	8. 25% of 64 is <u>24</u>. 18. 12 is <u>50</u>% of 6.
	9. 25% of 96 is <u>24</u>. 19. 12 is <u>100</u>% of 12.
	10. 10% of 200 is <u>20</u>. 20. 12 is <u>50</u>% of 24.

There are five mistakes on Pat's paper.

What percent of the problems on the test did Pat do wrong? _____

(speech bubble) 5 is what percent of 20?

How many of the problems did Pat not do? _____

What percent of the problems on the test did Pat not do? _____

What percent of the problems on the test did Pat do correctly? _____

How would you rate Pat's work? ☐ Poor   ☐ Fair   ☐ Good   ☐ Excellent

36

You know how to find these percents of some numbers:

1% 10% 25% 50% 100% 200% 300%

You can use these familiar percents to figure out other percents.

_____% = 25% + 1%   _____% = 50% − 10%

_____% = 10% + 10% + 1%   _____% = 50% − 1% − 1%

_____% = 25% − 1% − 1%   _____% = 100% − 10% + 1%

24% = _____   51% = _____

75% = _____   35% = _____

9% = _____   299% = _____

The first and second problems in each group are easy. You can find the answers to the two harder problems by adding or subtracting the first two answers.

25% of 200 is **50**.
1% of 200 is **2**.

| 24% of 200 is **48**. | 26% of 200 is **52**. |

50% of 600 is _____.
1% of 600 is _____.

| 51% of 600 is _____. | 49% of 600 is _____. |

25% of 160 is _____.
10% of 160 is _____.

| 35% of 160 is _____. | 15% of 160 is _____. |

200% of 36 is _____.
25% of 36 is _____.

| 175% of 36 is _____. | 225% of 36 is _____. |

100% of 210 is _____.
10% of 210 is _____.

| 90% of 210 is _____. | 110% of 210 is _____. |

10% of 400 is _____.
1% of 400 is _____.

| 11% of 400 is _____. | 9% of 400 is _____. |

10% of 500 is _____.

1% of 500 is _____.

| 11% of 500 is _____. | 9% of 500 is _____. |

100% of 40 is _____.

10% of 40 is _____.

| 90% of 40 is _____. | 110% of 40 is _____. |

200% of 60 is _____.

10% of 60 is _____.

| 190% of 60 is _____. | 210% of 60 is _____. |

100% of 36 is _____.

25% of 36 is _____.

| 75% of 36 is _____. | 125% of 36 is _____. |

10% of 800 is _____.
1% of 800 is _____.

11% of 800 is _____.

25% of 300 is _____.
1% of 300 is _____.

24% of 300 is _____.

___ % of 700 is _____.
___ % of 700 is _____.

9% of 700 is _____.

100% of 30 is _____.
10% of 30 is _____.

90% of 30 is _____.

100% of 24 is _____.
50% of 24 is _____.

150% of 24 is _____.

___ % of 40 is _____.
___ % of 40 is _____.

225% of 40 is _____.

26% of 200 is _____.

24% of 200 is _____.

24% of 800 is _____.

210% of 40 is _____.

190% of 40 is _____.

175% of 40 is _____.

35% of 80 is _____.

15% of 80 is _____.

15% of 20 is _____.

301% of 700 is _____.

299% of 700 is _____.

290% of 700 is _____.

38

Estimating Percents

Use a number in the box to name the shaded part of each circle.

23%	~~55%~~	5%	75%	33%	95%	$2\frac{1}{2}$%	$12\frac{1}{2}$%

55% ___ ___ ___ ___ ___ ___ ___

Match a number in the box with each approximate amount.

9%	26%	52%	47%	199%	105%

a little more than one quarter	a little more than one half	a little less than one half	a little less than one tenth	a little more than all	a little less than double
___	___	___	___	___	___

Shade part of each circle. Use a straightedge so that your work will be neat.

Shade about 50%.    Shade about 60%.    Shade about 70%.    Shade about 80%.

Shade about 25%.    Shade about 75%.    Shade about 10%. 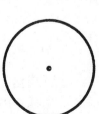   Shade about 5%.

Which set of percents best fits the pie chart? Write the percents on the pie pieces.

10%	90%
25%	75%
40%	60%
50%	50%

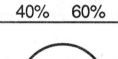

10%	90%
25%	75%
40%	60%
50%	50%

10%	90%
25%	75%
40%	60%
50%	50%

10%	20%	70%
20%	30%	50%
25%	25%	50%
30%	30%	40%

10%	20%	70%
20%	30%	50%
25%	25%	50%
30%	30%	40%

Divide each circle below to show the percents given. Write the percents on your pieces of pie. You'll have to estimate to decide how big to make each piece.

50%	50%

40%	60%

80%	20%

50%	40%	10%

In the year 2000, it is estimated that 6,121,000,000 people will live on earth.

Make a pie chart to show the information below. Label each piece of pie with the name of a continent and a percent. The size of each should match its percent.

About 60% will live in Asia.

About 15% will live in Africa.

About 11% will live in Europe.

About 9% will live in South America.

About 5% will live in North America.

World Population

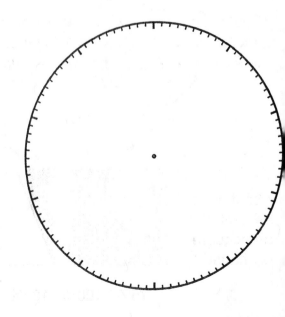

40

Circle the percent that best answers the question.

About what percent of the square is shaded?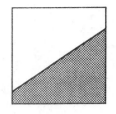

25% 50% 75%

About what percent of the dots are circled?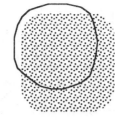

25% 50% 75%

About what percent of the circle is shaded?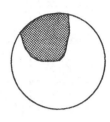

25% 50% 75%

About what percent of the cup is filled?

10% 25% 50%

About what percent of the square is shaded?

1% 10% 25%

About what percent of the dots are circled?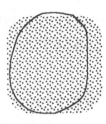

25% 50% 75%

About what percent of the square is shaded?

25% 50% 75%

About what percent of the circle is shaded?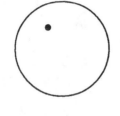

1% 10% 25%

About what percent of the wall is painted?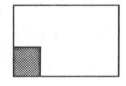

10% 25% 50%

About what percent of the square is shaded?

25% 50% 75%

About what percent of a cup is filled?

25% 50% 75%

About what percent of the rug is stained?

1% 10% 25%

About what percent of one square is shaded?

75% 100% 125%

About what percent of one square is shaded?

75% 100% 125%

Percent Is Based on 100

Pat plays on the school basketball team. In the first game of the season, Pat made 3 out of 4 shots. In the second game Pat made 4 out of 5 and in the third 7 out of 10. Pat wondered, "In which game did I shoot the best?" How can Pat find out?

It would be easy to tell if Pat had tried the same number of shots in each game. Let's pretend that Pat had continued to make 3 out of every 4, or 4 out of every 5, or 7 out of every 10 shots for 100 tries.

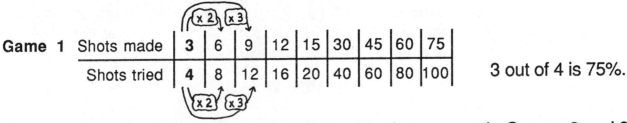

Game 1

Shots made	3	6	9	12	15	30	45	60	75
Shots tried	4	8	12	16	20	40	60	80	100

3 out of 4 is 75%.

You complete the tables below to find Pat's shooting percent in Games 2 and 3.

Game 2

Shots made	4								
Shots tried	5	10	20	30	40	50	100		

4 out of 5 is _____%.

Game 3

Shots made	7								
Shots tried	10	20	30	40	50	100			

7 out of 10 is _____%.

In which game did Pat shoot the best? _____

Complete the tables to find the shooting percents of three other players in Game 1.

Sandy

Shots made	13								
Shots tried	20	40	60	80	100				

13 out of 20 is _____%.

Terry

Shots made	6								
Shots tried	10	20	30	40	50	100			

6 out of 10 is _____%.

Chris

Shots made	12								
Shots tried	25	50	75	100					

12 out of 25 is _____%.

Who was the most accurate shooter, Sandy, Terry or Chris? _____

42

Complete the table to solve each problem below.

Jan got 12 out of 20 votes for class president. What percent of the votes did Jan get?

Votes for Jan	12				
Total votes	20	40	60	80	100

(x2) (x3)

12 out of 20 is _____%.
Jan got _____% of the votes.

Tanya got 18 hits in 40 times at bat playing softball last season. What was her batting average?

Hits	18				
Times at bat	40	20	100		

(÷2) (x5)

18 out of 40 is _____%.
Her batting average was _____%.

Philip got 24 out of 30 problems correct on his math test. What percent of the problems did he get correct?

Problems correct	24				
Problems on test	30	10	50	100	

(÷3) (x5) (x2)

24 out of 30 is _____%.
Philip got _____% of the problems correct.

The algebra class at Smalltown High School has 36 students. 30 is supposed to be the class size limit. What percent of the size limit was the algebra class?

Students in class	36				
Class size limit	30	10	50	100	

36 compared to 30 is _____%.
There were _____% as many students in the class as there were supposed to be.

Make up your own table to solve each problem below.

18 of 20 students in Ms. Serra's French class came to class last Friday. What percent of the class was present?

Present					
In the class					

18 out of 20 is _____%.
_____% of the class was present.

Pat and Chris are digging a ditch that will be 25 meters long. They have dug 7 meters. What percent of the ditch have they dug?

Meters dug					
Meters of ditch					

7 out of 25 is _____%.
They have dug _____% of the ditch.

The Bay High girl's softball team has won 18 of their last 30 games. What percent of their last 30 games have they won?

Games won					
Games played					

18 out of 30 is _____%. They have won _____% of their last 30 games.

Key to Percents Book 1

Practice Test

Name _____

Date _____

Write each percent in three ways.

Using words	As a fraction with denominator 100	As a number with percent notation
25 out of every 100		
	$\frac{5}{100}$	
		100%

What percent of each figure is shaded?

10 equal pieces

 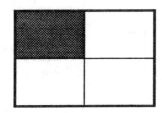

_____% is shaded.   _____% is shaded.   _____% is shaded.

_____% is not shaded.   _____% is not shaded.   _____% is not shaded.

Put hair on 50% of the heads.

Put a nose on 25% of the heads.

Put a smile on 100% of the heads.

0% of 35 is _____.   25% of 1600 is _____.

25% of 16 is _____.   0% of 7643 is _____.

50% of 48 is _____.   100% of 985 is _____.

100% of 80 is _____.   50% of 4444 is _____.

2 is _____% of 8.   6 is _____% of 12.

Fill in the missing percent. Then answer the question.

Students at Cabot High School

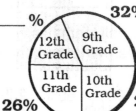

_____% 12th Grade, 9th Grade 32%, 10th Grade 25%, 11th Grade 26%

What percent of the students at Cabot High are in the 12th grade?

_____

44

Practice Test - Page 2

100% of 25 is _____.   10% of 600 is _____.   1% of 900 is _____.

200% of 25 is _____.   20% of 600 is _____.   2% of 900 is _____.

300% of 25 is _____.   30% of 600 is _____.   3% of 900 is _____.

Complete the sales receipt.

Buena Sports		
Item	Price	
1 Sweatshirt	12	95
1 Pair Sneakers	29	95
Subtotal		
10% Tax		
Total		

One quarter is _____% of a dollar.

A dime plus a nickel is _____% of a dollar.

1 centimeter is _____% of a meter.

50 centimeters is _____% of a meter.

One decade is _____% of a century.

10% of one hour is _____ minutes.

25% of 200 is _____.

1% of 200 is _____.

24% of 200 is _____.   26% of 200 is _____.

25% of 40 is _____.

10% of 40 is _____.

35% of 40 is _____.

Circle the percent that best names the part that is shaded.

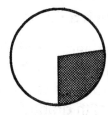

10%   25%   50%

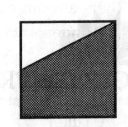

25%   50%   75%

Complete the table to answer the question.

Aran got 32 out of 40 problems correct on his math test. What percent did he get correct?

Problems correct	3 2				
Problems on test	4 0				

32 out of 40 is _____%.

Aran got _____% of the problems correct.

Key to Percents®

Book 1: *Percent Concepts*
Book 2: *Percents and Fractions*
Book 3: *Percents and Decimals*
Answers and Notes for Books 1–3
Reproducible Tests for Books 1–3

Also Available

Key to Fractions®
Key to Decimals®
Key to Algebra®
Key to Geometry®
Key to Measurement®
Key to Metric Measurement®

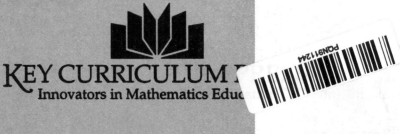

KEY CURRICULUM
Innovators in Mathematics Educ

ISBN 091368457-0

9 780913 684573